COOPERATIVE LEARNING
IN THE SCIENCE CLASSROOM

Author:
Linda Lundgren

GLENCOE

McGraw-Hill

New York, New York   Columbus, Ohio   Woodland Hills, California   Peoria, Illinois

Send all inquiries to:
Glencoe/McGraw-Hill
8787 Orion Place
Columbus, OH 43240

ISBN 0-02-826430-4

Printed in the United States of America.

17 18 19 20   066   04 03 02 01 00

TABLE OF CONTENTS

ACKNOWLEDGEMENTS

I would like to thank the teachers of the Jefferson County District in Colorado with whom I have consulted, especially Sally Ogden who has inspired many teachers to incorporate cooperative learning in their classrooms. A very special thank you goes to James Middleton, who has been my cooperative teammate as we developed plans, solved problems, and brainstormed cooperative learning ideas for biology students at Bear Creek High School.

I have relied a great deal on the works of many researchers, teachers, and cooperative learning trainers including Dee Dishon and Pat Wilson O'Leary, Ted and Nancy Graves, David and Roger Johnson, Spencer Kagan, Sally Ogden, Gene Stanford, Robert Slavin, and Jack Hassard.

I am appreciative of the support of Science and Technology Curriculum Director, Harold Pratt, and other administrators including my principal, Maran Doggett. Above all I wish to thank the many students who have taught me which cooperative learning strategies work well. These students are the true authors of this booklet.

INTRODUCTION

The effective use of cooperative skills is becoming increasingly necessary to cope successfully in today's team-oriented workplaces. Because of the increasing importance of cooperative interaction, it is essential that educational strategies include cooperative learning. It is not enough for science students to learn only subject matter. They also must learn skills for working relationships, such as how to listen, respond, agree, disagree, clarify, encourage, and evaluate. These skills are necessary for team members to work together productively.

This booklet presents jargon-free cooperative learning skills and strategies suitable for the science student. The number of cooperative learning terms will be minimized. Strategies suggested capitalize on the interests and strengths of students. Activities suggested involve their interest in how things work, their fascination for new and future technology, and their desire to manipulate materials. Included are suggestions for practicing the working relationship skills that students need. Examples of these skills are staying on task, dealing with distractions, and disagreeing in an agreeable way.

WHAT IS COOPERATIVE LEARNING?

Cooperative learning is an old idea. As early as the first century A.D., philosophers argued that in order to learn, one must have a learning partner. The basic elements of cooperative learning are as follows:

1. Students must perceive that they "sink or swim together."
2. Students are responsible for everyone else in the group, as well as for themselves, learning the assigned material.
3. Students must see that they all have the same goals.
4. Students must divide up the tasks and share the responsibilities equally among group members.
5. Students will be given one evaluation or reward that will apply to all members of the group.
6. Students share leadership while they acquire skills for collaborating during learning.
7. Students will be held individually accountable for material worked on in cooperative groups.

Differences between cooperative learning groups and traditional groups	
Cooperative learning groups	**Traditional groups**
Shared leadership	One leader
Positive interdependence	No interdependence
Heterogeneous membership	Homogeneous membership
Instruction in cooperative skills	Assumption of effective social skills
Responsibility for all group members' achievement	Responsibility for individual achievement
Emphasis on task and cooperative relationship	Emphasis only on task
Support by teacher	Direction by teacher
One group product	Individual products
Group evaluation	Individual evaluations

WHY USE COOPERATIVE LEARNING?

Studies show that cooperative learning techniques promote more learning than competitive or individual learning experiences. Increased learning occurs regardless of student age, subject matter, or learning activity. Complex learning tasks such as problem solving, critical thinking, and conceptual learning improve noticeably when cooperative strategies are used. Students often consider learning to be complete once they have mastered lists of facts. However, they are more likely to use higher levels of thinking during and after discussions in cooperative groups than when they work competitively or individually. Thus, students retain learned material for a longer period of time.

Studies show that in classroom settings, adolescents learn more from each other about subject matter than they do from the teacher. Consequently, the development of effective communication should not be left to chance. The cooperative learning method capitalizes on students' propensity for interaction. Research also indicates that cooperative learning has an extremely positive impact on students who are low achievers.

The work of scientists involves debating ideas in the light of new evidence and rational criticism. Many experts in education agree that science classes should realistically reflect the actual work of scientists. Cooperative learning is especially appropriate in science classes.

Common misconceptions about cooperative learning are held by students, teachers, and parents. Many students fear that the work will not be divided evenly, that one person will have to do all the work. Parents worry that their child's grades will go down. Some teachers assume that all group work is cooperative. These misconceptions and others will be dispelled as students, parents, and teachers become familiar with cooperative learning strategies.

Research-supported benefits of cooperative learning for low-achieving students

- Increased time on task
- Higher self-esteem
- Improved attitude toward science and school
- Improved attendance
- Lower drop-out rate
- Greater acceptance of individual differences
- Less disruptive behavior
- Reduced interpersonal conflict
- Less apathy
- Deeper comprehension
- Greater motivation
- Higher achievement
- Longer retention
- Increased kindness, sensitivity, and tolerance

Misconceptions and truths about cooperative learning	
Misconception	**Truth**
All group work is cooperative.	Students must use specific skills in order to work cooperatively.
Students can work cooperatively to complete individual assignments.	The work of a cooperative group results in a joint product or group conclusions.
One student in a group usually ends up doing most of the work.	Use of cooperative skills ensures an equal division of the task.
Students perceive group evaluations as being unfair.	Once they experience cooperative learning, most students acknowledge that group grades are fair.
Competition is a more realistic strategy.	Almost all human activity is cooperative.
Cooperative learning strategies can enliven a classroom when they are used occasionally.	Effective cooperative learning evolves as it is used over a period of time.
Difficult classroom management problems accompany the use of cooperative learning strategies.	More management problems may exist in traditional classrooms, which require silence and attention.
High achievers suffer academically when they work in heterogeneous cooperative groups.	Research shows that high achievers in cooperative learning situations do as well as or better than their peers in traditional classrooms.
All group members do the same work at the same rate.	Students in the same group can be assigned different tasks.
Cooperative learning is easy to implement.	The concept of cooperative learning is simple, but its implementation is complex.

GETTING STARTED WITH COOPERATIVE LEARNING IN SCIENCE

Arrange Your Room

Students in groups should face each other as they work together. It is helpful to number the tables so you can refer to the groups by number. Or, have each group choose a name.

Decide on the Size of the Group

Groups work best when composed of two to five students. The materials available might dictate the size of the group. If attendance is a problem, groups may need to redistribute frequently. If you begin with activities that can be completed in one class period, students will not be faced with new group members in the middle of an activity because of absences.

Assign Students to Groups

Groups should reflect life. The group should be mixed socially, racially, ethnically, by gender, and by learning abilities. If students are allowed to choose their own groups, there may be less task-oriented behavior; and the homogeneity that usually results will not allow them the opportunity to hear views that may differ from their own. Heterogeneity mirrors the real world, which encompasses encountering, accepting, and appreciating differences. If a group finds that things are not working out, members may complain that they want to switch groups. If students are told that they can change groups after they have proven their ability to work effectively for a given time, maybe two weeks, they often decide they don't want to switch.

Random group assignments have an advantage because students perceive them as being fair. However, randomly selected teams should not stay together for more than a couple of activities if the teams turn out to be homogeneous.

Change Groups Periodically

Some teachers keep their groups together for a week, a quarter, or a teaching unit. The only criterion would be to have groups stay together long enough to experience success as a group. The advantage of changing groups is that students will have more opportunity to deal with a variety of classmates.

Prepare Students for Cooperation

This is a critical step. Tell students about the rationale, procedures, and expected outcomes of this method of instruction. Students need to know that you are not trying to force them to be friends with other people, but are asking them to develop working relationships with people who come together for a specific purpose. Point out to your students that they will need to do this later on in life. Explain that using cooperative skills might feel awkward at first, like any new experience, such as throwing a football or playing a guitar for the first time. Stress that it will take time to learn the necessary cooperative skills. Explain the three basic rules of cooperative learning:

1. Stay with your group.

2. Ask a question of your group first, before you ask your teacher.

3. Offer feedback on ideas; avoid criticizing people.

You may want to post these rules in your room. Pay close attention to the first rule until students become comfortable with their groups.

If cooperative learning has not been used at your school, you may wish to explain the method and use the **Sample Letter to the Principal** on page 29, **Sample Letter to Parent or Guardian** on page 30, and **Sample Letter to the Substitute Teacher** on page 30.

Make Use of This Booklet

Start small. You do not have to incorporate all of this information into your planning for your first cooperative groups.

There are three levels of cooperative skills and strategies: beginning, intermediate, and advanced. Students should master many of the basic skills before they practice intermediate, and then advanced skills and strategies. Cooperative learning skills and strategies are listed on pages 15–26.

Have Students Get Acquainted

Students who know each other are more comfortable working together. In addition to the ones suggested on pages 15–17, many of the references in the bibliography on page 27 have activities students can do to learn names and get to know their classmates. Have students do several "getting to know you" activities to become comfortable with students they do not know.

Explain the Day's Lesson

Make a reusable transparency of the **Action Plan** on page 31, or write the headings below on the board at the start of each activity. Fill in the **Action Plan** by following the guidelines below:

Form Groups: Number of students in each group
Topic of the Day: General academic topic
Task: Work that students would complete, such as a worksheet or model
Goal: The academic objective
Cooperative Learning: The targeted cooperative learning skill or strategy
Special Instructions: Method of evaluation or other information specific to the activity

THE TEACHER'S ROLE

Comparison of teacher's role using cooperative and traditional learning methods	
Cooperative learning	**Traditional learning**
Supports	Directs
Redirects questions	Answers questions
Teaches social skills	Makes rules
Manages conflicts	Disciplines
Structures interdependence	Encourages independence
Helps students evaluate group work	Evaluates individuals
Structures controversies	Directs recitation and discussion
Provides resources	Serves as major resource

Introduce the Cooperative Skill or Strategy for the Lesson

Review the **Action Plan.** Explain how many students should be in each group and how they should be formed; the topic, task, and goal for the day; and the cooperative learning skill or strategy to be used. The explanation of the cooperative skill or strategy must involve discussion by students, rather than simply explanation by the teacher. In a discussion of the questions listed below, the students take ownership of the process of cooperative learning and see that it meets their needs.

When introducing a cooperative learning skill, ask the following questions:
1. *What is _____ ?* Fill in the blank with the skill or strategy you are introducing.
2. *In any group, why is it important to use _____ ?* You may fill in the blank with the skill, such as using agreement, paraphrasing, or active listening, that you are introducing.
3. *How will the working relationships in your group improve if you use this skill or strategy?* If time permits, you may want students to discuss each of the above questions in their groups before calling for an answer.
See pages 22–26 for a variety of skills in the format of questions and possible answers.

Use T-Charts

Draw a large "T" on the board. Write the skill on top of the "T," the words, "sounds like"

on the left side of the "T," and the words "looks like" on the right side of the "T." Ask students to suggest phrases that they could say to each other as they practice a particular skill and to suggest body language that communicates this skill. See pages 38–40 for T-chart suggestions. Use this information to give students hints or to get them started thinking of their own phrases.

Monitor Student Use of Skills and Strategies

Just because students have been placed in groups and given directions does not mean that they will automatically work cooperatively. When you first use cooperative learning, you will want to observe very closely to see that things are off to a good start. Move about the room and listen to make sure students realize that their communications matter and that you are aware of their progress. Use one of the formal observation forms provided on pages 36–37. Record the number of times you observe the use of skills expected on that particular assignment for each group. Do not try to count too many different behaviors at the beginning. Occasionally you may want to try having a student make the group observations.

Provide Assistance with the Task

Clarify instructions, review concepts, or answer questions, but avoid interfering in the groups' process. It is critically important to let groups make decisions on their own. Students must be allowed to make mistakes, then evaluate themselves to see where they went wrong. Because of their past experiences, students who see you hovering nearby will automatically start asking questions. Your first response should be, "Have you asked everyone in your group?" You may want to ask groups to appoint one student from each group to be the "Spokesperson" or the "Liaison" between you and the group. That group member would be the only one allowed to speak with you from the group.

Your role will be supportive supervisor rather than direct supervisor. For example, as a supportive supervisor, you will help a group that has gotten stuck and is experiencing a high level of frustration. You might do this by asking a few open-ended questions. In a conflict situation, you might ask the group members to come up with some strategies for handling the conflict.

Intervene to Teach Cooperative Skills

If you observe that some groups have more problems than others with learning cooperative skills, you may wish to intervene by asking group members to think about why they are not being effective and asking them to come up with a solution. If they decide on an inappropriate solution, such as ignoring the problem, explain that it is unacceptable. You might also ask groups to refer to relevant T-charts that are displayed around the room.

Provide Closure for the Lesson

Students should be asked to summarize what they have learned and be able to relate it to what they have previously studied. You may want to review the main points and ask students to give examples and answer final questions.

Evaluate the Group Process

In order for groups to be aware of their progress in learning to work together, they must be given time to evaluate how they are working together. Give students a few minutes at the end of the lesson to decide if they achieved the criteria you set up for that particular lesson. Everyone in the group should participate in the evaluation. Sometimes you may incorporate their evaluation into the overall grade for the activity. Student evaluation of their own success with cooperative strategies and skills is essential for progress to be made. It is easy for students to continue the science task until the end of class and neglect the group evaluation. If this occurs on a regular basis, your groups will not make progress with cooperative learning. Another problem that may occur in students' evaluations is that the answers are vague. Responses such as "We cooperated" or "We did OK" are too superficial to be beneficial to the group. The following is a suggested list of evaluation questions to ask at the end of any cooperative learning lesson. They may be written on the board or directly on the laboratory directions or worksheets students are using during the activity. Choose one or two questions each time students use cooperative learning. Most of the time you will want students to do one evaluation of the group as a whole. Occasionally it is beneficial to have students reflect on their individual contributions to the group. (See page 35.)

Suggested responses to specific group behaviors during cooperative work	
Group behavior	**Suggested teacher response**
Use of cooperative skill	Encourage with specific compliments.
Quick completion of task	Ask questions that require deeper analysis.
Off task	Have group suggest methods to keep members on task.
Questioning of teacher	Ask, "Have you asked everyone in your group?"
Difficulty with the resolution of a problem	First ask, "What have you done so far to resolve the problem?" Then ask, "What could you do next?"

Evaluate Student Academic Learning

Cooperative Tests and Quizzes

Even though students work together, individuals are accountable for their own learning. Group grading should not be done until students are comfortable with cooperative learning. Evaluation tools should include traditional tests and quizzes.

There are several ways to determine the group score for students who have worked together to prepare for a quiz or test—average all group members' scores, grade only one group member's test, or have one student from a group take the test. You select the student to take the test on the day the test is administered. Students should not know ahead of time who will be chosen. All other group members would get the grade produced by the test taker.

Some experts suggest basing the group score on the improvement of individual scores. In this way, others are not penalized if a group member's entering achievement level is low. To receive maximum credit, all group members should score higher than they did on the last test.

Another way to administer points on an individual test would be to award extra credit points to everyone in a group if all group members achieve a predetermined score on the test. For example, if everyone in the group achieves above 90 percent, the group would get bonus points. If everyone in the group achieves an 80 percent or better, group members get half of the bonus points awarded to those achieving 90 percent.

An oral quiz can be given using the following format. You ask a question. Group members consult on the answer. You choose a student to answer. However, at this point no help can be given by the rest of the group.

Another method of recognizing group accomplishment on individual tests is to award group or class improvement points based on the following scoring system: 3 points if everyone in the group scores 10 or more above their base or average, 2 points if everyone in the group scores 5–9 above their base, and 1 if everyone in the group scores 4 below to 4 above their base. The group receives the sum of its members' individual improvement points. Using this system, the group members who improve the most will have great value to the group. Often these students are ones who have low motivation and skills and as a result of their new importance, develop higher self esteem and work habits. The improvement points can be recorded on a large bar graph for individual groups, or in a thermometer-type graph for the entire class.

What to do with bonus points is an issue on which experts take different stands. Some teachers like competition between teams; others feel that competition is opposed to the principles of cooperative learning. Also in question is the subject of material rewards for points. Some teachers feel that payment for learning will lower the motivation and pleasure of learning for its own sake. Most experts agree that a limited use of rewards enhances the classroom experience. Rewards are suggested on page 12.

For a change of pace, give a group test. Tell each student to prepare one 3×5 note card of information. Allow group members to share note cards during the test. Have students who do not prepare a note card take the test individually. Group tests should not be used frequently, as individual mastery and learning is the ultimate goal.

Cooperative Homework

Group members divide up a worksheet of questions or chapter review questions and receive a grade based on their combined answers. Another alternative is for students to answer the same set of questions, compare their answers the next day and submit one group set of the best answers. All papers must be attached to the group set in order to receive credit. Another alternative is to award bonus points if all group members complete their homework.

Cooperative Lab Reports, other Written Reports, and Worksheets

Students prepare individual reports or worksheets while consulting with each other and comparing results and conclusions. One paper is collected from the group to be the score for all group members. The group does not know prior to collection whose paper it will be, though the method of collection is announced at the beginning of the activity. Another method involves dividing up a report or worksheet into sections. Group members put together the sections for a combined report.

Cooperative Projects: Skit, Group Presentation, Model, Art or Craft Work, Debate, and other Group Products

In order to provide individual assessment of group projects, you may want to combine a group score with an individual score based on effort and participation. You might ask group members to numerically evaluate the contributions of members and justify the scores. Another method of assessing the individual contribution to a group project is to give the group the combined total of all their scores and allow the group to divide them up fairly among themselves.

Students do well evaluating their own group products, such as reports and presentations, if they have a list of grading criteria. An advantage is that students perceive grades they assign themselves as being fair. You may wish to involve the students in establishing the grading criteria. This method also helps them perceive the process as being fair.

Use a Cooperative Contract

After group members have been working together for some time they will learn each other's strengths and weaknesses. You may want them to fill out the **Student Agreement** on page 34 and award points for achieving contracted goals. To implement the use of a contract, first ask students to list what their group does well. Tell them to be specific. Also ask them to list the specific things they need to do to make their group work better. They should list specific behaviors for specific people, but focus on the behavior that needs to be changed, not the person. Beside each item they should indicate how they each could help in making the change. Discuss each group's contract with them and tell them how many points you will award if they can "fix things" by the selected date. You have the option to give groups with bigger problems more points. Ask students to evaluate themselves on a scale of 1–10 each day on their progress toward their goal.

Keep in mind that cooperative learning doesn't just happen. The first few days may seem like bedlam, with some students upset, others mistrustful, and others off task. Both you and your students will make mistakes. Just be patient and keep at it. Research shows that it takes three years for a teacher to become unconsciously skillful using a new teaching method. However, your students will benefit even during the first lesson.

Give Commendations and Awards

Most students respond positively when recognized for achievement. Some teachers prefer not to use tangible material rewards, while others find them highly effective. Suggestions for recognition and reward are given in the table on the next page. Many businesses are willing to provide free passes or coupons. In addition, you may want to use a certificate such as the one on page 34.

Methods to recognize and reward achievement in academic and working relationship skills		
Recognition	**Privileges**	**Rewards**
Certificate	Free time	Free passes: dance, sports event, movie, skating rink
Smile	Excuse from homework	
Pat on back	Early lunch	Treat for class: movie, pizza
Display of work	Music during class	Gift certificates: fast food, hair styling
Standing ovation	Time for game playing	
Message to parents	Library time	Good grade
Quarterly recognition ceremony	Computer time	Points earned toward: prize, popcorn party, free pass to athletic event
Personal note from teacher to student	Use lab equipment	
List of group achievement points on a posted chart	Help other students or teachers	
List of class achievement points on a posted chart		
Students of the week		

TROUBLESHOOTING

A Student Resists Working in a Group

In the rare case of determined opposition to working in a group, you may want to allow the student to work individually. The student may eventually reconsider. Alternatively, suggest that the student commit to a group for a limited time, perhaps three weeks. Monitor the group closely during this time. Encourage group members to offer possible solutions to the problem.

A Student Behaves Inappropriately

Whenever possible, allow the group to deal with the problem. By intervening you give up your most powerful tool, peer influence. You also risk sending the message that the students are not capable of solving their own problems. When necessary, offer assistance in the form of specially designed group analysis questions, talk to the offender after class, or use the following procedure:

• Ask, "What have you done so far to resolve the problem?"
• Ask, "What might you do next?" and have the group brainstorm solutions.
• Approve an appropriate plan of action.
• Have group members write down and sign their plan. Have them refer to it daily during group work.
• Consider giving a daily reward for progress toward the goal.
• Consider giving extra credit points when the problem is resolved.

Students Do Not Effectively Use Cooperative Skills

Structure lessons so that groups can identify their weak skills and practice them. Allow ample time for groups to evaluate their work using the Group Analysis questions. Encourage individuals or groups to commit to improving specific skills by signing a contract such as the one shown on page 34. Recognize and reward improvement. Keep in mind that giving praise or encouragement may be very difficult for students who have a reputation for being tough.

Group Members Do Not Share Equally in the Work

Divide up materials so that each group member has information others need. Give the group only one worksheet. Assign each group member an essential role.

Give each group member five slips of paper. For each contribution to the group task, a student must relinquish a slip of paper. When a student has no more slips of paper, he or she may no longer contribute.

If one group member consistently does not participate despite efforts by others to include him or her, take this into consideration when assigning the group grade so other group members are not penalized.

A Student's Ability Is Considerably Lower Than That of Other Group Members

Tailor the weaker student's task. Provide appropriate reference materials for him or her to use. You may want to adapt tests and quizzes or the scoring method you use, or automatically add points to the student's individual grade when using it to figure the group grade.

A Student Is Absent

Have the group suggest appropriate make-up work. Approve the assignment.

Consider combining groups if two or more students are absent from a group. Another option is to have floaters who are academically successful and skilled in using cooperative strategies fill in empty places.

A Student Is Chronically Absent

Assign the student as an extra member to a group with a core that is usually present. Or, have the student fill in for absent students when he or she does come to class. Offer a permanent assignment when attendance improves.

Students Continue to Use "Put-downs," Ridicule, and Demeaning Remarks

Ask the group to make a list of all the positive qualities they can think of for each group member. Ask them to make a written plan of action for dealing with their negative communication. Ask them what would be a good reward if they are able to improve in one week.

A Student Is Extremely Shy

Use team building, trust building, and active listening activities which create an atmosphere of acceptance and respect for each other. Make complimenting, encouraging participation, and appreciating individual differences the cooperative skills that groups must practice. Begin with shy students in a smaller group. Ask the group to take on task roles and assign the shy student the role of reader, recorder, or spokesperson.

A Student Is a High Achiever

Reward him or her for helping others. Ask him or her to work with a difficult partner and give the group a bonus for the difficult partner's success. Assign challenging roles that the student does not usually take. This student may do observations of the cooperative efforts of the entire class. Group the high achievers together occasionally to work on an especially fast-paced, challenging project. If necessary, reassure the student (and parent if necessary) that research shows that mastery and retention of academic material by high-ability students is found to be higher in cooperative than in competitive or individualistic learning situations.

A Student Actively Attempts to Sabotage Group Work or Products

Reinforce daily any behavior that is near the cooperative goal. Assign a cooperative skill tied to the disruptive behavior. Write the skill on an overhead transparency beside the student's name. Tally the number of times the cooperative skill is used by the student, rewarding positive behavior at the end of class. Choose a reward this student would like. Tell the group that they will receive this reward when they earn a certain number of points for taking positive steps to correct the behavior. Use a Student Contract as described on page 34. Role play the problem with other students in the class and have a class discussion about how to solve the problem. As a last resort, ask the disruptive student to work alone until he or she is willing to practice cooperative skills.

The Noise Level Rises Too High

Develop a signal that means "quiet." You may simply raise your hand, with students following your example as soon as they see you raise your hand. It may be a quick flick of the light switch or a bell. Have students practice the cooperative skill of using "six-inch voices" early in the year and again if noise is a problem. Assign the role of "noise monitor" to one member of each group. Educate your colleagues and principal about the difference between "noise" and the "beehive of activity" involved in cooperative learning. Reward groups for keeping the noise level down.

Group Consistently Refuses to Work with a Particular Student

Give the outcast student roles with leadership responsibility. Use careful strategies for grouping. Be sure one student in the group has some positive feelings about the outcast. Each day ask the group to start by saying one positive comment to each person in the group. Use strategies for conflict resolution and structure the activities so the student is needed by the group for them to be successful. Strategies such as jigsaw and limiting materials and information may be included. Ask the group to practice skills such as honoring individual differences and showing appreciation and empathy.

COOPERATIVE LEARNING STRATEGIES

The activities and strategies that follow are structured ways of introducing cooperative learning skills. Working relationship skills help groups function effectively by developing communication and interpersonal skills. There are two kinds of roles—working relationship roles that help students with communication and interpersonal relationships and task roles that help students get their work done (see table on page 16). If working relationships are good, the work is accomplished effectively. Begin with the first four working relationship roles in the table and change or add as needed for your class or for individual groups.

The cooperative strategies on the following pages are divided into Getting Acquainted, Comparing Cooperative Learning with Individual Learning, Building Team Identity, Beginning Strategies, Intermediate Strategies, and Advanced Strategies. These cooperative strategies should be introduced as described in the first two parts of **The Teacher's Role**, page 9. The skills that appear on pages 22–26 are in the question and answer format discussed in **The Teacher's Role**. Examples of **T-charts** are given on pages 38–40 for many of the skills.

Students should master many of the beginning skills (not necessarily in the order presented) before going on to intermediate and advanced skills. Some classes may need to practice certain skills over and over. They may never get beyond beginning skills for an entire year. You should not view this as a problem or failure. It is not wise to push students into using skills that are too advanced for them. Those classes with previous experience may be comfortable with more advanced skills early in the year.

Getting Acquainted

In order for groups to work together effectively it is important that students become comfortable with each other. Icebreakers, games, and activities that encourage getting acquainted should be done before introducing skills and strategies. The books in the bibliography on page 27 provide many additional suggestions.

Working relationship roles	Task roles
Taskmaster: Encourages group members to stay on task and keeps track of time.	**Timekeeper:** Keeps track of time to be sure the task is completed in allotted time.
Cheerleader: Gives positive recognition and encouragement to group members.	**Recorder:** Writes ideas or answers on the group paper.
Summarizer: Periodically tells the group what it has accomplished and what it still needs to do. Records ideas and answers.	**Materials handler:** Obtains needed materials for lab work.
Coordinator: With group input, divides up the task. Reminds group members of their responsibilities. Is also the spokesperson, the only group member who communicates with the teacher.	**Cleanup crew:** Puts away materials and cleans workspace.
	Refuse manager: For activity requiring a great deal of cleanup, coordinates cleanup crew.
Checker: Checks to be sure that everyone understands and agrees with answers.	**Keyboard operator:** For computer work, does the typing.
Coach: Encourages group members to perform their tasks.	**Monitor watcher:** For computer work, checks for accuracy.
Energizer: Enlivens a group that has become apathetic.	**Technical support crew:** For lab or computer work, performs necessary tasks not specifically mentioned.
Clarifier: Explains details, expectations, and directions.	**Liaison:** The only person from a group who speaks with the teacher.
Paraphraser: Repeats what others have said in his or her own words.	**Reader:** Reads directions or other materials so the group can understand them.
Gatekeeper: Makes certain that each person participates and that no one person dominates the group.	
Quiet captain: Keeps noise level down.	
Opinion seeker: Asks questions of group members.	
Prober: Asks questions that go beyond the scope of the lesson.	

Interviews

Make up a list of statements for which students must get signatures of classmates. The statement list is prefaced by the announcement *Find someone who . . .*
Examples: has the same color eyes, has same favorite TV program; went to the same place on vacation

Introductions

Give a set amount of time for the groups to go through a list of questions about each group member. Teacher picks up student sheet and asks other group members questions from the sheet.

Examples: favorite color, hobby, sport; best thing that ever happened to them

Categories

Entire class breaks up into groups as you announce, "Get together with everyone who has the same favorite animal as you." As soon as students are in groups announce another category.

The Lineup

Ask your students to line up by specific measurements such as the number of letters in their names, shoe size, hair length, or the farthest away from home they have been.

Human Scavenger Hunt

Give students a list of 10 characteristics. Students are to identify another student having any of the 10 characteristics. A signature is needed beside each characteristic.

Matching

Students match cards of paired items. Each student has a card and must find the student who matches. Items could be vocabulary words and meanings, foods and the animals that eat them, diseases and their treatments.

Survival

Tell students that each group represents survivors of an arctic plane crash, a desert plane crash, or shipwrecked on an island. Give each group a list of items that must be ranked in order of importance to their survival.

One-line drawing

Give each student in a group a different color marker and tell them that they may not speak. Have them pass around a large piece of paper on which they add one line at a time per student to make a drawing. There should be no discussion prior to making the drawing. After the time is up, they should discuss their drawing and think about how it reflects the group.

Comparing Cooperative Learning with Individual Learning

It is essential that students be given the opportunity to see for themselves the benefits of cooperative learning as you introduce this learning strategy. Ask students to work for five to ten minutes on something a bit too difficult for them to do alone. It could be identification of animal tracks, a list of health-related myths and facts which they must identify as true or false, a complex crime scene which must be resolved, or "mind bender"-type puzzles. Ask them how it felt when they could not solve the problem. Ask them to work in a group of four to resolve the same problem. Ask them to compare individual learning with cooperative learning.

Building Team Identity

After students have worked in their groups a few times, it is beneficial to do an activity that will help develop positive group identity and give members a sense of unity. The strategies listed below will serve this purpose and also relate to academic content. Ask students to create one of the following that will reflect an important characteristic of their group as well as an important feature of the current unit of study: *Team Slogan, Team Banner, Team Logo, Team Cheer, Team Mural, Team Name.*

Beginning Cooperative Strategies

Focus

Before a lab, lecture, film, or new unit, ask students to write down everything they already know about the topic. Afterward have them discuss the new knowledge they gained.

Study Buddies

In groups of two, have each member question the other about material being studied. Tell them they will receive bonus points if they score above a certain percentage.

Checkmates

Have groups compare homework answers or class worksheet answers. They should discuss answers which differ and come to agreement on the best answers and change them. Collect one paper per group. Do not tell them in advance which paper you are planning to collect.

Turn to Your Neighbor

After a lecture, an explanation of directions, or an assignment, ask students to turn to their neighbor and have each point out or summarize a key point of what was said.

Jigsaw

On a reading assignment of one or two pages that does not have sequential importance, divide up the reading among the members of a group of three or four. Each person reads his or her part of the assignment and then teaches it to the other group members. Other group members should be quizzed by the teacher to make sure that they understand the material.

Do not divide up entire chapters or units until students have developed expertise in cooperative learning.

Brainstorming

Use this strategy to generate a large number of alternative ideas for discussion of a question. Ask students to make a group list. They should not evaluate the ideas until the list is complete. Encourage them to build on each other's ideas and go on side tracks and into weird and silly places. All ideas are acceptable for inclusion on a brainstorm list. Evaluation begins when no one has further ideas or the time is up.

Blackboard Share

Ask one group member from each group to put their best idea or answer on the board. This strategy allows the groups still at work to consider the ideas on the board as well as their own and perhaps come to an even higher level of thinking.

Write a Note

All members of a team of four write a note that begins, "What I understand about this lesson or chapter is. . . . I am still having trouble with. . . ." Ask them to trade notes with someone who is not having the same trouble and reply to the note. They should write the note as if they were writing a "real" note to a friend. Have them fold and pass the notes in the style currently in fashion in the school.

Advanced Cooperative Learning Strategies

After simple cooperative learning strategies are used, students will be ready to use more complex strategies. The strategies near the end of this list should be used after students have experience with earlier strategies.

Group Interdependence

Prepare for each group a set of 4 to 5 "clue cards" including a distractor for one concept. For example, you could make a set of clues for the properties of gases, another for solids, another for liquids. The sets are traded from one group to another as they figure them out. Use the following rules:

a. Do not allow anyone to see your clue.
b. Verbally communicate your clue to your group.
c. Decide which of the clues are distractors.
d. Decide what concept the clues represent.

Alternatives Grid

Prepare a worksheet as shown in Table 1 below. Ask students to practice a skill such as acknowledgement or disagreeing in an agreeable way as they fill in the grid for a topic that is controversial in nature.
Examples: endangered species, health issues, energy, gene therapy

Peer Feedback

This strategy is suitable for use with written assignments such as reports, essays, and lab

TABLE 1

Consequences	Alternative 1	Alternative 2	Alternative 3
Environmental			
Societal			

TABLE 2

Vocabulary words	Function	Looks like	Works like

reports. Make an evaluation checklist of content items and writing style items that you want students to check. Ask students to trade papers and evaluate items on the checklist on a scale of 1–5. Students should also indicate the best feature of the report and what could be improved. Encourage them to comment on what is good before saying anything negative. To avoid arguments, tell students to paraphrase comments so the person knows he or she has been understood.

Vocabulary Comparisons

Make a worksheet as shown in Table 2 on page 18. Have students name the real function of the word and give an object in everyday life it may resemble. Then indicate a process in everyday life that has a similar function. Students should brainstorm their ideas while using cooperative skills.

Example: In cells, the nucleus looks like a ball and works like a brain.

Yarn Talk

Give each group a ball of yarn at the beginning of a seat activity requiring discussion. The first speaker should tie the yarn loosely around his or her wrist. The speaker should pass the yarn to the person who speaks next. The second speaker winds the yarn loosely around his or her wrist and passes it on to the next speaker. At the end of the activity, have each group analyze its communication pattern.

Talking Chips

Give each group member seven small pieces of paper. Each time someone speaks, he or she must give up a piece of paper. When a group member is out of paper, he or she can no longer speak until everyone has used all their pieces of paper. This strategy will keep a group member from dominating the discussion.

Response Chips

Give each student a sheet of paper with the following list of responses. Each student should cut out the responses and place the appropriate one in the middle of the group each time he or she speaks. The response list should include two "give idea", two "restate speaker," two

"ask question," two "keep group on task," one "give encouragement," one "summarize progress," one "respond to idea," and other "chips" you may need in order to deal with specific problems.

Snowballing

A pair of students answer worksheet questions or compare lab report conclusions or other written work. Two pairs come together to review and compare answers. Two groups of four come together and compare. One person from a group of eight writes answers or conclusions on the board.

Roundtable

One group member has a pencil and paper. He or she reads a question out loud. Group members consult and refer to the textbook in order to agree on the answer. The group member who has the pencil and paper writes the answer. The answer sheet is passed to the next group member. Repeat the process until all questions are answered. One person in each group should check the answers using a key provided by the teacher. This strategy is especially appropriate for review questions.

Numbered Heads Together

Students in each group of four count off from one to four. Ask a question and tell the students how much time (about 30 seconds) they have to confer with group members. Students put their heads together in their groups to answer the question. Randomly call a number from one to four. Students with that number raise their hands if they think they can answer the question. There is no more talking after the number is called. If the student answers correctly, his or her group receives a point. If not, a student with the same number from another group may respond. This strategy is especially appropriate for review.

Group Directed Numbered Heads Together

This strategy is similar to *Numbered Heads Together*, except that groups direct the process. Student 1 in the first group asks a question and tells the other groups how much time they have to confer. Students in the other groups put their heads together to answer the question. Student 2 in the first group randomly calls a number from one through four. Students with that number raise their hands if they think they can answer the question. There is no more talking after the number is called. Student 3 in the first group calls on a student whose hand is raised to answer the question. Student 4 evaluates the answer. Groups take turns asking one question at a time.

Teammates Consulting

Everyone should have a worksheet. All pencils should remain in the middle of the table while group members read and discuss a question. When the group reaches agreement, everyone should pick up a pencil and write the answer. Collect only one worksheet per group. This strategy is an excellent method of reviewing or answering worksheet or lab questions.

Consulting Between Groups

Ask groups to trade papers with the members of another group. The group should analyze and comment, using cooperative skills, on answers with which they disagree.

Groups Visit

Three students from each group take their completed work and "visit" another group. One student in each group remains and presents his or her group's work to the visitors. The visitors compare their work and note any differences. Students return to their groups. A different group member then remains while the other three visit different groups. Visits continue until every student has visited three times and explained once. This strategy is useful for checking work.

Paraphrase First

Each time a group member has contributed an idea, another group member must correctly restate the idea before another idea can be contributed.

Making Analogies

Ask students to brainstorm ways a concept is like something in everyday life. This strategy works especially well when the scientific concept has a sequence to it. For example, students may compare performing an experiment and producing a play. Analogy ideas should be brainstormed by the group. These ideas might be sequential processes such as production of a newspaper, baking a cake, or taking a vacation. Give each group a large sheet of butcher paper and markers. Have each group make a labeled diagram of its analogy. The diagram should show the everyday process with vocabulary word labels for the scientific process. This strategy works well to help students understand a difficult scientific concept.

Planning a Lesson

Ask students to prepare a demonstration explaining a science concept for younger students.

Dramatization

Ask students to prepare and present a skit about the life of a famous scientist or scientific discovery. Prepare a checklist of items for them to consider as they prepare:
a. What will the setting be? Props?
b. What will each character say and do?
c. What important events will be portrayed?
d. How many scenes will be needed? What will happen in each scene?

Expert Groups

Divide a chapter or unit so that each student has responsibility for a section. Students from each group who are assigned the same section get together in "expert groups" to plan the best methods for teaching their section to the rest of their home group, review all the important details, and get help from the teacher about difficult details. The "experts" go back to their home group to teach the materials to them. On a larger scale, the home group may become the "expert group" for the entire class in making presentations for the whole class. The expert group should quiz the class or their home group so there is individual accountability for learning.

Jigsaw

Divide a chapter or unit so that each group member is responsible for a section. Each member makes a teaching plan consisting of objectives, concepts to be presented, notes, diagrams, and questions to test understanding for his or her section. Jigsaw is similar to Expert Groups but more difficult because low achieving students do not have an opportunity to consult with others as they do in Expert Groups.

STAD

The title of this strategy is an abbreviation for "Student Teams Achievement Division." Tell each group its base score or average to date. Divide students into pairs and have them divide a set of review questions among the pairs. Partners ask each other questions and determine answers. Share the questions and answers with the rest of the group after checking with an answer key. Partners should continue asking and answering questions until everyone can answer all the questions. Give an individual test that includes many of the questions that students studied together. Bonus points are given to groups whose average scores on the test are higher than their base scores.

Making a Group Contract

Ask each group to make a list of specific behaviors that can be changed to improve their group. "We need to work together better" is not specific. Examples of specific behaviors include: "Mary should come to class on time." and "Joe should stop reading car magazines." Beside each item, have students write how the change will be accomplished. The teacher will read the contract and write on it the number of points or other rewards the group will gain by meeting their contract in an allotted time period. Groups with more problems can be awarded more points. At the end of the time period each group must write a justification for the number of points they think they should receive.

Viewpoints

Select a societal issue for which students could take on roles in a debate within their groups. For example, in a debate about the tropical rain forest, one student could be an environmental scientist; another, an agricultural specialist; another a trade developer; and another, a human rights advocate for local citizens. Prepare a role card for each student. The card explains the viewpoint of each person.

Group Writing

Students take turns writing two sentences, reading out loud as he or she writes, and accepting suggestions from the group. After two sentences are written, the paper is passed to the next group member who will add two sentences. This strategy will work for a variety of writing assignments, such as summaries of science news articles, essays, science stories, biographies of famous scientists, and lab reports.

Energizing

Students create a cartoon, song, poem, or rap illustrating a concept related to the unit of study. Discuss some of the elements of humor, such as exaggeration, substitution, and using opposites that go into humor. Show cartoons that satirize topics being studied. Have examples of light poetry such as jingles and limericks. Suggest to students that they use science vocabulary from the unit of study to write new words to old tunes such as "Home on the Range."

Using Mnemonics

Ask students to develop mnemonic devices to help them remember sequences, pathways, and concepts. Ask them to develop mnemonic sentences in which each word begins with the same letter as a word to be memorized. For example, "King Philip came over for good spaghetti" is a mnemonic for the taxonomic hierarchy: Kingdom, Phylum, Class, Order, Family, Genus, Species. A mnemonic sentence is usually unrelated to the words to be memorized. It can be clever or unusual.

Debate

Pairs of students in a group of four are assigned either pro or con positions on a societal issue such as use of animals for research, euthanasia, or alternative energy sources. Students must defend the viewpoint assigned regardless of their own personal views on the matter. Students must compromise in their group as they debate the issue. Rules during debate are as follows: Criticize ideas, not people. Criticize to learn more about a position, not to judge it. Paraphrase others' statements. Ask for explanations. Modify your position in view of convincing arguments.

Group Decision Making

Give students a societal issue which must be resolved. Ask them to brainstorm a list of possible options and then rank them in their order of importance. They must use the following strategies to rank the options:

a. Develop a list of criteria to use in evaluating options.

b. Rate each option on a scale of 1–10.

c. Consider the pros and cons of each option. Tell the group they may not flip a coin, compromise, vote, eliminate extremes and choose the least controversial option, or allow one member to decide. Emphasize that the strategies just mentioned would be appropriate

under certain conditions, but that you want them to practice another strategy.

Coming to Consensus

Ask each group member to take a viewpoint on a societal issue. Prepare role cards as in *"Viewpoints"*. Use the following procedure based on the pipe smoking ceremony of Native Americans:

a. A group member holds a small tree branch and presents one argument for his or her viewpoint.

b. The tree branch is passed clockwise. The group member who receives it may speak or pass.

c. A person may speak only while holding the branch.

d. Speak to state opinions or to support previous statements, but not to disagree.

e. Continue passing around the branch until one group member makes a summary statement and all the other group members pass to indicate that they agree.

INTRODUCING COOPERATIVE SKILLS: QUESTIONS AND ANSWERS

Beginning Cooperative Skills

Using Agreement

1. What is meant by using agreement? Having the same opinion.

2. Why is it important to use agreement in your group? Group members will know who has the same opinions.

3. How will the working relationships in your group improve if you use this skill? Group members will feel that their opinions are worthwhile and important.

Acknowledging Contributions

1. What does acknowledging contributions mean? To notice or recognize what another group member has said or done. It is not necessarily the same as agreement. In fact, it may be criticism. Criticism should always be of ideas, not people.

2. Why is it important to acknowledge contributions? Group members realize that they are understood.

3. How will the working relationships in your group improve if you use this skill? Group members will feel that their opinions are worthwhile and important.

Using Quiet Voices

1. What does using quiet voices mean? Using "six-inch voices" that cannot be heard beyond the table.

2. Why is it important to use quiet voices? To be able to hear the conversation in your group.

3. How will the working relationships in your group improve if you use this skill? Group members will be able to hear clearly and not become frustrated by a loud classroom.

Taking Turns and Sharing

1. What does taking turns and sharing mean? Alternating who does specific tasks and takes specific responsibilities in the group.

2. Why is it important to take turns and share? Work will be done more effectively if all group members contribute in an organized manner.

3. **How will the working relationships in your group improve if you use this skill?** Group members will develop a sense of being a team working toward the same goal.

Staying with Your Group
1. **What does staying with your group mean?** Remaining in your seat or at your group work station.
2. **Why is it important to stay with your group?** Work will not be efficient if group members are wandering around the class-room.
3. **How will the working relationships in your group improve if you use this skill?** Groups that stay together will be able to depend on each other.

Staying on Task
1. **What does staying on task mean?** Continuing with assigned activity in spite of distractions.
2. **Why is it important to stay on task?** Activity will be accomplished on time with better accuracy and more opportunity for creativity.
3. **How will the working relationships in your group improve if you use this skill?** The group will be proud of their increased effectiveness in preparing assignments well.

Encouraging Participation
1. **What does encouraging participation mean?** Motivating all group members to contribute.
2. **Why is it important to encourage participation?** If one or two people do not participate or they contribute very little, the group product may not be finished in time, or it may be lacking in originality and imagination.
3. **How will the working relationships in your group improve if you use this skill?** Group members will feel that their contributions are important.

Inviting Others to Talk
1. **What does inviting others to talk mean?** Requesting that others speak.
2. **Why is it important to invite others to talk?** If one or two people do not participate or they contribute very little, the group product may not be finished in time or it may be lacking in originality and imagination.
3. **How will the working relationships in your group improve if you use this skill?** Group members will feel that their contributions are important.

Finishing on Time
1. **What does finishing on time mean?** Completing the activity by a designated time.
2. **Why is it important to finish on time?** Learning will be incomplete and unfinished work will receive lower grades.
3. **How will the working relationships in your group improve if you use this skill?** Group members will gain a sense of accomplishment and team spirit if they complete assignments and do them well.

Using Names and Looking at the Speaker
1. **What does using names and looking at the speaker mean?** Calling each other by name and using eye contact.
2. **Why is it important to use names and look at the speaker?** Everyone likes to be called by name. When you look at the speaker, he or she knows that you are paying attention.
3. **How will the working relationships in your group improve if you use this skill?** Group members gain a sense that they are important contributing members when names are used and eye contact is made.

Dealing with Distractions
1. **What does dealing with distractions mean?** Avoiding problems that result from diversions or inattention to assigned work.
2. **Why is it important to deal with distractions?** Distractions can keep a group from completing the assigned learning task.
3. **How will the working relationships in your group improve if you use this skill?** If positive steps are taken by groups to keep distractions at bay, they will develop a sense of accomplishment and maturity about mastering this skill.

Helping Without Giving Answers

1. **What does helping without giving answers mean?** Giving assistance without revealing the solution.
2. **Why is it important to help without giving answers?** If you give answers to other group members, they may not gain a sense of understanding of the concept.
3. **How will the working relationships in your group improve if you use this skill?** As all group members contribute ideas to the solution of a problem, they will all gain a sense of accomplishment and sense of pride in their group.

Honoring Individual Differences

1. **What does honoring individual differences mean?** Being respectful of all students' unique cultures, life experiences, and ethnicities.
2. **Why is it important to honor individual differences?** Hostility can be avoided. Group harmony can be promoted.
3. **How will the working relationships in your group improve if you use this skill?** Tensions will be reduced. A sense of belonging and camaraderie may develop. Individuals may develop qualities of kindness, sensitivity, and tolerance.

Intermediate Cooperative Skills

Showing Appreciation and Empathy

1. **What is meant by showing appreciation and empathy?** Showing respect, understanding, and sensitivity to opinions that are different from yours.
2. **Why is it important to show appreciation and empathy?** Hostility can be avoided. Group harmony can be promoted.
3. **How will the working relationships in your group improve if you use this skill?** Tensions will be reduced. A sense of belonging and camaraderie may develop. Individuals may develop qualities of kindness, sensitivity, and tolerance.

Using "I" Messages

1. **What are "I" messages?** Telling how you feel by using the word "I" when you speak. For example, rather than saying, "You are wrong," say "I don't think so."
2. **Why is important to use "I" messages?** When you use the first person to express yourself, others do not feel threatened or accused so hostility can be avoided.
3. **How will the working relationships in your group improve if you use this skill or strategy?** Tensions will be avoided and group members will feel that they are appreciated.

Disagreeing in an Agreeable Way

1. **What is disagreeing in an agreeable way?** Expressing a different opinion or answering in a courteous and good-natured way.
2. **Why is it important to disagree in an agreeable way?** Personal criticisms and "put downs" create a negative atmosphere in the group.
3. **How will the working relationships in your group improve if you use this skill?** When ideas, not people, are criticized, group members do not feel insulted and hostilities can be avoided.

Active Listening

1. **What is active listening?** Using physical and verbal messages to let the speaker know that you are energetically processing information.
2. **Why is it important to use active listening?** Understanding of concepts will be enhanced and the group product will reflect a high level of thinking and communication.
3. **How will the working relationships in your group improve if you use this skill?** When the speaker is not interrupted and all students are attentive to communication, group members feel that what they are contributing is worthwhile.

Questioning

1. **What does questioning mean?** Inquiring, or asking for more information or clarification.
2. **Why is it important to use questioning?** Concepts can be clarified; someone off task can be encouraged to participate; shy group members can be motivated to get involved.
3. **How will the working relationships in your**

group improve if you use this skill?
Communication will be enhanced.

Summarizing
1. **What is summarizing?** Reviewing information.
2. **Why is it important to use summarizing?** It helps organize what has been done and what still needs to be done.
3. **How will the working relationships in your group improve if you use this skill?** When group work is completed effectively and efficiently students are proud of their group.

Paraphrasing
1. **What is paraphrasing?** Restating information in other words.
2. **Why is it important to use paraphrasing?** Information can be clarified and points emphasized.
3. **How will the working relationships in your group improve if you use this skill?** Communication is enhanced.

Managing and Organizing
1. **What does managing and organizing mean?** Planning and arranging work so it will be accomplished efficiently and effectively.
2. **Why is it important to manage and organize?** Assignments will be completed efficiently and effectively.
3. **How will the working relationships in your group improve if you use this skill?** Goals are more easily achieved when a group is well organized. Accomplishing goals will give groups pride in their ability to work together and also generate positive attitudes.

Checking for Accuracy
1. **What does checking for accuracy mean?** Comparing answers, making certain that answers are correct.
2. **Why is important to check for accuracy?** Work will be free of mistakes or errors. A better understanding of subject matter will develop.
3. **How will the working relationships in your group improve if you use this skill?**

Improved group products will foster positive interdependence.

Taking Responsibility
1. **What does taking responsibility mean?** Being willing and able to assume duties and obligations for yourself and for the group to accomplish the assigned work.
2. **Why is it important to take responsibility?** Assignment cannot be completed if group members do not take their responsibilities seriously.
3. **How will the working relationships in your group improve if you use this skill?** Groups members who take responsibility for their own learning as well as for the learning of the group learn more than they could alone.

Using Patience
1. **What does using patience mean?** Being tolerant of others, sticking to work in spite of difficulties, not making hasty decisions.
2. **Why is it important to use patience?** There will be fewer frustrations, less tension, and less stress created for group members.
3. **How will the working relationships in your group improve if you use this skill?** Group members will feel accepted, feel a sense of accomplishment for sticking to the task, and will develop a sense of maturity.

Keeping Calm/ Reducing Tension
1. **What does keeping calm/reducing tension mean?** Promoting a peaceful group atmosphere.
2. **Why is it important to keep calm and reduce tension?** A tranquil atmosphere in the group promotes higher levels of learning.
3. **How will the working relationships in your group improve if you use this skill?** Hostility will be controlled. No one feels threatened or insulted when tensions are reduced.

Advanced Cooperative Skills
Elaborating
1. **What does elaborating mean?** Expanding on concepts, conclusions, and ideas related to the topic.

2. Why is it important to elaborate? Deeper comprehension and higher achievement will result.

3. How will the working relationships in your group improve if you use this skill? Higher achievement will result in greater motivation and improved attitudes.

Probing

1. What does probing mean? Asking questions that take the subject to greater depth to ensure the correct answer. Nonaccusatory questions such as "Why?" and "Can you give an example?" are used.

2. Why is it important to use probing? Ensures that answers are correct.

3. How will the working relationships in your group improve if you use this skill? Higher achievement will lead to higher self-esteem.

Asking for Justification

1. What does asking for justification mean? Proving that the answer is right or giving a reason for the answer.

2. Why is it important to ask for justification? Helps students think through answers and be more certain of their accuracy.

3. How will the working relationships in your group improve if you use this skill? Higher achievement will lead to positive attitudes.

Advocating a Position

1. What does advocating a position mean? Taking a stand on a question or issue.

2. Why is it important to be able to advocate a position? In order to sway people to your line of thinking, it is important to be non-judgmental and respectful of others' views while presenting your position positively.

3. How will the working relationships in your group improve if you use this skill? Respect for others' opinions will reduce conflict in the group.

Setting Goals

1. What does setting goals mean? Establishing priorities.

2. Why is it important to set goals? Work is done more efficiently if goals are defined.

3. How will the working relationships in your group improve if you use this skill? There will be less disruptive and distracting behavior.

Compromising

1. What does compromising mean? Settling an issue by mutual agreement.

2. Why is it important to compromise? Compromise builds respect for others and reduces interpersonal conflict.

3. How will the working relationships in your group improve if you use this skill or strategy? Learning to criticize ideas and not people, to paraphrase others' statements to ensure understanding, and modifying your position in view of convincing arguments will lead to maturity and development of good judgment.

Confronting Specific Problems

1. What does confronting specific problems mean? Addressing problems with "I" messages; not accusing, calling names, or using sarcasm; addressing only behavior that can be changed, not personality traits or disabilities; aiming to resolve a problem, rather than winning.

2. Why is it important to confront specific problems? Interpersonal conflict will be reduced and qualities of kindness, sensitivity, and tolerance will develop.

3. How will the working relationships in your group improve if you use this skill? Tension will be avoided and interpersonal relationships will improve. The group will function better as a team and accomplish goals more effectively.

BIBLIOGRAPHY

Adams, Dennis M., and Mary E. Hamm. *Cooperative Learning, Critical Thinking and Collaboration Across the Curriculum.* Springfield, IL: Charles C. Thomas Publisher, 1990.

Aronson, Elliot, Nancy Blaney, Stephan Cookie, Jev Sikes, and Matthew Snapp. *The Jigsaw Classroom.* Beverly Hills: Sage Publications, 1978.

Clarke, Judy, Ron Wideman, and Susan Eadie. *Together We Learn.* Scarborough, Ontario: Prentice-Hall, 1990.

Cohen, Elizabeth G. *Designing Groupwork: Strategies for the Heterogeneous Classrooms.* New York: Teachers College Press, 1986.

Dishon, Dee, and Pat Wilson O'Leary. *A Guidebook for Cooperative Learning: A Technique for Creating More Effective Schools.* Holmes Beach, FL: Learning Publications, Inc., 1987.

Gibbs, Jeanne. *Tribes: A Process for Social Development and Cooperative Learning.* Santa Rosa, CA: Center Source Publications, 1987.

Graves, Nancy, and Ted Graves. *What Is Cooperative Learning? Tips for Teachers and Trainers.* Santa Cruz, CA: Cooperative College of California, 1990.

Hassard, Jack. *Science Experiences.* Menlo Park, CA: Addison-Wesley, 1990.

Johnson, David W. *Reaching Out: Interpersonal Effectiveness and Self-Actualization.* Englewood Cliffs, NJ: Prentice-Hall, 1986.

Johnson, David W., and Frank P. Johnson. *Joining Together: Group Theory and Group Skills.* Englewood Cliffs, NJ: Prentice-Hall, 1987.

Johnson, David W., and Roger T. Johnson. *Circles of Learning: Cooperation in the Classroom.* Alexandria, VA: Association for Supervision and Curriculum Development, 1984.

Johnson, David W., and Roger T. Johnson, *Creative Conflict.* Edina, MN: Interaction Book Company, 1987.

Johnson, David W., and Roger T. Johnson. *Cooperation and Competition: Theory and Research.* Edina, MN: Interaction Book Company. 1989.

Johnson, David W., and Roger T. Johnson. *Learning Together and Alone: Cooperative, Competitive, and Individualistic Learning.* Englewood Cliffs, NJ: Prentice-Hall, 1990.

Johnson, David W., Roger T. Johnson and Edythe Johnson Holubec. *Cooperation in the Classroom.* Edina, MN: Interaction Book Company, 1990.

Johnson, Roger T., David W. Johnson, and Edythe Johnson Holubec. *Structuring Cooperative Learning Lesson Plans for Teachers.* Edina, MN: Interaction Book Company, 1987.

Kagan, Spencer. *Cooperative Learning Resources for Teachers.* San Juan Capistrano, CA: Resources for Teachers, 1989.

Ogden, Sally, Cindy Hrebar, and Jim Fay. *A Guide to Improving Teacher Effectiveness in a Cooperative Classroom.* Unpublished Manuscript, 1984.

Sharan, Shlomo, Editor. *Cooperative Learning, Theory and Research.* New York: Praeger, 1990.

Slavin, Robert E. *Cooperative Learning: Student Teams.* Washington, D.C.: National Education Association, 1987.

Slavin, Robert E. *Student Team Learning: An Overview and Practical Guide.* Washington, D.C.: National Education Association, 1988.

Stanford, Gene. *Developing Effective Classroom Groups: A Practical Guide for Teachers.* New York: A & W Publishers, 1977.

SAMPLE LETTER TO THE PRINCIPAL

Dear

This year I will be implementing cooperative learning in science classes. I trust that this will meet with your approval. You may know that cooperative learning is a thoroughly tested method of instruction in which students work in small groups toward a common goal while using specific cooperative skills.

In cooperative learning, members of heterogeneous groups share leadership, have positive interdependence, are responsible for each other's learning, produce a group product, and receive group rewards. The emphasis is not only on the biological task, but also on good working relationships and communication.

Educational research shows the following benefits of students engaged in cooperative learning:

• Greater academic achievement
• Higher self-esteem
• Use of higher-level thought processes
• Increased time on task

Cooperative learning employs highly structured planning and clear goals for individuals as well as groups. Research shows that students learn a significant amount of subject matter from each other. Rather than leave successful communication to chance, the cooperative learning method capitalizes on the importance of interpersonal relationships to adolescents. Students will be taught cooperative skills to ensure productive group work. Group products will be evaluated, but individuals will remain accountable for their own learning.

There are many worthwhile benefits of cooperative learning in any classroom setting, but current research shows that the low achiever has even more to gain, such as

• Improved attitude toward science and school
• Improved attendance
• Lower drop-out rate
• Reduced interpersonal conflict
• Greater motivation

I would like you to visit my classes periodically and discuss your observations with me as the year progresses. I shall keep you informed of our progress.

Sincerely,

SAMPLE LETTER TO PARENT OR GUARDIAN

Dear Parent or Guardian:

This year I will be implementing cooperative learning techniques in science class along with traditional teaching methods. Cooperative learning is a thoroughly tested method of instruction in which students work in small groups toward a common academic goal while using specific cooperative skills.

Educational research shows the following benefits for students engaged in cooperative learning:

- Greater academic achievement.
- Higher self-esteem.
- Increased motivation to learn.
- Use of higher-level thought processes.

Cooperative learning employs highly structured planning and clear goals for individuals as well as groups. Even though group products are evaluated, individuals are accountable for their learning.

Please contact me if you have comments or questions. I am looking forward to a successful and exciting year of science.

Sincerely,

NOTE TO THE SUBSTITUTE TEACHER

Students in my classes work in cooperative learning groups. This means that they are responsible not only for their own learning, but for that of other group members as well. They "sink or swim together." They divide up the task and share responsibilities equally. One evaluation is used when group products are produced; but students are also held accountable for their own learning as individuals.

Please use the attached transparency to explain the day's lesson. Review Procedure on the student worksheet. As students work, encourage them to use the cooperative skill or strategy for the lesson. When they ask you a question, ask them if they have asked everyone in their group before you give an answer. You may want to use the attached observation sheet to record students' use of cooperative skills as they work together.

For a brief overview of cooperative learning, please read pages 5–7 in the booklet.

FORM GROUPS:

TOPIC OF THE DAY:

TASK:

GOAL:

COOPERATIVE LEARNING:

SPECIAL INSTRUCTIONS:

STUDENT QUESTIONS FOR EVALUATION OF COOPERATIVE LEARNING

1. Evaluate your group on a scale of 1–10, with 10 being best, on your use of _____ today. Justify your evaluation.

2. Evaluate each group member on a scale of 1–10, with 10 being best, on your use of _____ today. Justify your evaluation by giving specific examples.

3. **a.** List two main strengths of your group.

   b. List the most important thing your group could do to improve. Be specific. You cannot list something like "cooperate more." That statement is not specific.

   c. List what your group could do to accomplish the improvement listed above. Be specific. You cannot list vague suggestions such as "We should work together better."

4. **a.** What was the effect of using _____ today? (agreement, paraphrasing, or other skill or strategy)

   b. Why is this skill or strategy important?

5. What did you do today that helped improve the working relationships in your group? Give specific examples.

6. When you were using _____ today, what specific words did you use?

7. Did you _____ today? If not, why not? Be specific.

8. The next time you use _____ , how could you improve? Give specific suggestions.

9. **a.** What part of this cooperative learning skill or strategy was easy? Why?

   b. What part of this cooperative learning skill or strategy was difficult? Why?

10. Did you succeed in carrying out the cooperative learning strategy today? Why or why not? Give specific reasons.

11. What are the advantages of using _____ ? (Use this question when you have used a particular strategy such as roundtable or jigsaw.)

12. What kind of nonverbal communication did you use to show _____ ?

13. Describe a situation in real life in which you could use the skill you practiced today.

14. **a.** What did you like best about using the cooperative learning skill or strategy for today? Why?

    b. What did you like least about using the cooperative learning skill or strategy for today? Why?

15. Compare the working relationships your group has developed to date with the working relationships in your last group. Be specific.

16. Explain in specific detail how your group could have been more effective today.

17. **a.** What have you done so far to resolve the problems in your group? Be specific.

    b. What could you do next? Be specific.

18. **a.** What happened in your group today that made you feel good?

    b. What didn't you like about what happened in your group today?

19. **a.** What cooperative skill does your group use most effectively?

    b. What cooperative skill does your group need to improve? How will you accomplish this improvement? Give specific suggestions.

20. Approximately how many times did group members use _____ ? What one thing could you do to become better at _____ ?

21. How does your group deal with distractions and interruptions? Give specific examples.

22. **a.** How did individual group members contribute to the group's working well today? Describe specific examples of behavior rather than focusing on the person.

    b. How did individual group members

cause problems for the group today? Describe specific examples of behavior rather than focusing on the person.

23. Circle the following phrases that best describe your group:

Got started immediately

Stayed on task

Worked well together

Did not get started immediately

Did not stay on task

Did not cooperate

24. Circle the following statements that best describe your group:

Everyone contributed.

We helped each other.

We listened carefully.

Not everyone contributed.

We did not help each other enough.

We did not listen carefully.

25. Circle the statements that best describe your feelings:

I am satisfied with my group.

My group works productively.

I am not satisfied with my group.

My group does not work productively.

26. Describe the ways that your group helped each other learn today.

27. Describe something that people in your group learned that you could not have learned working alone.

28. a. What did your group accomplish today?

b. What helped you get it done?

c. What got in your way?

29. a. What one word would you use to describe your group's working relationships today?

b. What one word would describe the way you would like your group to be?

Occasionally, such as before students change to new groups and after they have been together for several weeks, have groups answer some or all the **Monthly Group Evaluation** questions on page 35.

COOPERATION AWARD

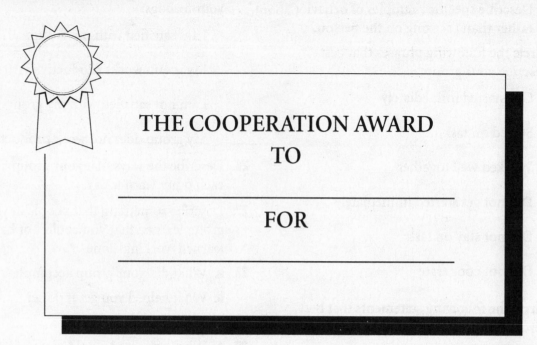

THE COOPERATION AWARD

TO

_____

FOR

_____

STUDENT AGREEMENT

I/We will improve in the cooperative skill of

_____

(skill)

by _____.

(date)

Signed:_____

Date: _____

MONTHLY INDIVIDUAL EVALUATION

This form should be used once per quarter, or about every two months.
Check the appropriate space.

	Always	Usually	Sometimes	Never
1. I contribute ideas and information.	_____	_____	_____	_____
2. I encourage others.	_____	_____	_____	_____
3. I ask questions	_____	_____	_____	_____
4. I ask for help when I need it.	_____	_____	_____	_____
5. I help other group members learn.	_____	_____	_____	_____
6. I help keep the group on task.	_____	_____	_____	_____
7. I make sure everyone in the group understands.	_____	_____	_____	_____
8. I use "I" messages.	_____	_____	_____	_____
9. I am positive.	_____	_____	_____	_____
10. I listen to the teacher's instructions.	_____	_____	_____	_____

11. Two skills I need to improve are _____ and _____ .

MONTHLY GROUP EVALUATION

This form should be used once per quarter, or about every two months.
Check the appropriate space.

	Always	Usually	Sometimes	Never
1. Everyone contributes.	_____	_____	_____	_____
2. We encourage each other.	_____	_____	_____	_____
3. We ask each other questions.	_____	_____	_____	_____
4. We share the work equally.	_____	_____	_____	_____
5. We help each other learn.	_____	_____	_____	_____
6. We stay on task.	_____	_____	_____	_____
7. We solve our group's problems.	_____	_____	_____	_____
8. We use "I" messages.	_____	_____	_____	_____
9. We are positive.	_____	_____	_____	_____
10. We listen to the teacher's instructions.	_____	_____	_____	_____

11. Two skills we need to improve are _____ and _____ .

TEACHER OBSERVATION FORM

Form 1

Beginning Observations

Start by teaching one skill and recording each use of the skill with a mark. Later you may want to record students' initials and observe more skills at the same time.

Cooperative skill	Group							
	1	2	3	4	5	6	7	8

TEACHER OBSERVATION FORM

Form 2

Group Member Observation Form   Group # _____

Start by teaching one skill and recording each use of the skill by a group member with a mark. Later you may want to observe more skills at one time. Or, keep the sheets and add on new skills later.

Cooperative skill	Group member				

T-CHARTS

Active listening

Sounds like	Looks like
Say "uh-huh" as the speaker talks.	Nod
Use acknowledgement from other T-chart.	Eye contact
	Lean forward
	Wait through pauses
Use open-ended questions to keep the speaker talking.	Smile
	Relaxed posture
	Hands unclenched
Paraphrase what the speaker says.	Arms not crossed
Use encouragement to keep the speaker talking.	
Do not immediately disagree and turn the focus on your own opinion.	
Summarize the speaker's statement.	

Asking for justification

Sounds like	Looks like
How do you know that?	Point at work
Please explain that in another way.	Eye contact
	Lean forward in chair
What is your reasoning?	Questioning expression
How did you get that result?	Interested expression

Paraphrasing

Sounds like	Looks like
Let's get on task.	Smile
So, what you mean is . . .	Eye contact
In other words . . .	Hand gesture: turn palm up and open
To sum up . . .	Lean forward in chair
You mean that . . .	
Let me get this straight . . .	

Dealing with distractions

Sounds like	Looks like
Could we have everyone's attention?	Hand gesture to get attention
	Raised eyebrows
Could we wait on that?	Point to center of table
We could use your help.	Eye contact
Hello?	Look at group members
Excuse me.	Tap shoulder
We'll talk about that later.	
I want to get this done first.	
Time's running out.	
I have to help my group.	

Encouraging participation

Sounds like	Looks like
Use person's name.	Smile
Ask for feedback:	Nod
What do you think?	Eye contact
How are you doing so far?	Lean forward in chair
Is this O.K. with you?	Hand gesture: "Come on"
What part doesn't make sense?	
What's wrong?	
What can I do to help?	
We need your ideas.	
Encouragement	
Are we all together on this?	

Showing appreciation and empathy	
Sounds like	**Looks like**
I see what you mean. Hey, thanks! I appreciate that. That's great! I see your point. Yes. I hear you. That must have made you feel . . .	Smile Eye contact Nod head Lean forward in seat Look of concern Touch arm

Staying on task	
Sounds like	**Looks like**

Let's get on task.
Let's do that later.
Come on.
Let's get going.
We need you.
Let's get with it.
Let's socialize at
  lunch.
Time's running out.
	Smile
	Nod
	Hand gesture:
	  "Come on"
	Point to work
	Tap on shoulder to
	  get attention
	Eye contact

Questioning	
Sounds like	**Looks like**

Is this O.K. with you?
Would you help me
  with this?
Do you think it should
  be like this?
What's next?
Can you explain this
  another way?
What do you think?
How does that work?
How can we fix this?
	Smile
	Raised eyebrows
	Both palms up and
	  out to your sides
	Puzzled frown
	Eyes open wide

Acknowledging contributions	
Sounds like	**Looks like**

Let's get on task.
I see.
I didn't think of that.
That's a possibility.
I understand.
O.K.
That makes sense.
That's interesting.
Thanks, that helps.
That's reasonable.
	Smile
	Eye contact
	Lean forward in
	  chair
	Nod
	Pat on the back
	Relaxed posture
	  with hands not
	  clenched,
	  arms not crossed

Using "I" messages	
Sounds like	**Looks like**

I think that will work.
I think that won't
  work.
I don't want to do it
  that way.
I'm unhappy with
  that.
I don't think that's
  right.
I feel that we
  should . . .
I like that.
	Eye contact
	Lean forward in chair
	Nod
	Relaxed position
	Hands unclenched
	Arms not crossed

Using agreement	
Sounds like	**Looks like**

I agree.
I like that idea.
All right!
Wow!
Right on!
Great!
Good idea!
Super job!
You're right!
Far out!
Absolutely!
Yes!
	Nod
	Smile
	Eye contact
	Shake hands
	Clap
	Pat on the back
	Hand gestures:
	  Thumbs up
	  O.K.
	  "Give me five"
	  High five

Inviting others to talk	
Sounds like	**Looks like**

We need your help.
Could you get more
  materials?
How could you help?
What do you think?
What should we do
  next?
Would you help us
  with this?
Encouragement
Use person's name.
It's your turn to . . .
I'm glad you're
  in our group
  because . . .
	Smile
	Eye contact
	Lean forward in chair
	Pat on the back
	Nod
	Relaxed posture
	Hands unclenched
	Arms uncrossed

Probing	
Sounds like	**Looks like**

Are you sure about
  that?
How is that like. . . .?
Why?
Can you give me
  another example?
Can you take that a
  bit further?
What if . . . ?
Can we look at that
  another way?
Let's think about how
  that fits with what
  we did last week.
	Puzzled frown
	Lean forward in chair
	"The thinker"
	  position: elbow on
	  desk, chin resting
	  on hand
	Eye contact

Disagreeing in an agreeable way	
Sounds like	**Looks like**

I hear what you're
  saying, but what
  about . . . ?
O.K. or Yes, I under-
  stand, but . . .
Your point is well
  taken, but . . .
That's a good point,
  however . . .
I think our views
  differ on that.
Would you agree
  with me to the
  point of . . .
	Slight smile
	Calm facial
	  expression
	Lean forward in seat
	Relaxed upper body
	Eye contact
	Intent listening
	  behavior

Keeping calm and reducing tension	
Sounds like	**Looks like**

Let's be positive.
Let's cool it.
Could we lower our
  voices?
Let's start over.
Could we all calm
  down a little,
  please?
First let's decide what
  you two do agree
  on.
Would you go along
  with this, instead?
	Eye contact
	Concerned facial
	  expression
	Hand out, open,
	  palm up
	Slight smile